LET'S LOOK AT LANDFORMS

THAT'S A MOUNTAIN!

BY DWAYNE HICKS

Gareth Stevens PUBLISHING

Please visit our website, www.garethstevens.com. For a free color catalog of all our high-quality books, call toll free 1-800-542-2595 or fax 1-877-542-2596.

Library of Congress Cataloging-in-Publication Data
Names: Hicks, Dwayne, author.
Title: That's a mountain! / Dwayne Hicks.
Description: New York : Gareth Stevens Publishing, 2022. | Series: Let's look at landforms | Includes index.
Identifiers: LCCN 2020032278 (print) | LCCN 2020032279 (ebook) | ISBN 9781538263716 (library binding) | ISBN 9781538263693 (paperback) | ISBN 9781538263709 (set) | ISBN 9781538263723 (ebook)
Subjects: LCSH: Mountains–Juvenile literature.
Classification: LCC GB512 .H53 2022 (print) | LCC GB512 (ebook) | DDC 551.43/2–dc23
LC record available at https://lccn.loc.gov/2020032278
LC ebook record available at https://lccn.loc.gov/2020032279

Published in 2022 by
Gareth Stevens Publishing
29 E. 21st Street
New York, NY 10010

Designer: Tanya Dellaccio
Editor: Greg Roza

Photo credits: Cover John Philip Harper/Cultura/Getty Images; pp. 3–22 (background texture) MicroOne/Shutterstock.com; p. 5 George Rose/Getty Images News/Getty Images; p. 7 saiko3p/iStock/Getty Images Plus/Getty Images; p. 9 Moelyn Photos/Moment/Getty Images; p. 11 Arterra/Universal Images Group/Getty Images; p. 13 Jim Sugar/Corbis Documentary/Getty Images; p. 15 PapaBear/iStock/Getty Images Plus/Getty Images; p. 17 donwinni/Shutterstock.com; p. 19 Takeshi.K/Moment/Getty Images; p. 21 Bloomberg/Getty Images.

Printed in the United States of America

CPSIA compliance information: Batch #CSGS22: For further information contact Gareth Stevens, New York, New York at 1-800-542-2595.

Find us on  

CONTENTS

Boldface words appear in the glossary.

What's a Mountain?

A mountain is a raised area of land taller than a hill. Mountains are the biggest landforms on Earth! Earth has 109 mountains that are more than 4.47 miles (7.2 km) tall! They are all in Asia.

Hills and Peaks

People around the world have different ideas about how big a landform needs to be to call it a mountain. In some areas, mountains may look like tall hills. In other areas, mountains have huge **cliffs** and snowy **peaks**.

Mountain Ranges

Mountains often form in a line called a range. Ranges can be thousands of miles long! When a mountain range forms in the ocean, the mountain peaks may rise above the ocean's **surface**. This can create a **chain** of islands.

Mountain Forces

Earth's surface is not one big piece. It's made of smaller "plates" that move very slowly. When two plates bump into each other, the land can raise up and form a mountain. This process takes many, many years.

Volcanoes!

Volcanoes form when openings in Earth's surface let **lava** pour over the land. The lava cools and hardens. This process forms mountains! Some islands are actually giant volcanoes on the ocean floor. The Hawaiian Islands were all formed by volcanoes.

The Catskills

Some mountains form because of **erosion**. For example, the Catskill Mountains in New York State were once a large, flat area, called a plateau. Over time, wind and water wore away at the flat land. This created the Catskills.

Landslide!

Erosion can weaken a mountain over many years. Cracks in the rock form when water freezes and unfreezes. Rock can fall away from the mountain. When a large amount of land falls away, it's called a landslide.

Green Hills, Snowy Tops

Tall mountains have different weather at different **elevations**. Low areas often have forests and green hills. Tall peaks can be covered with snow all year. The tree line on a mountain is the line above which no trees can grow.

People and Mountains

People have lived on and around mountains for a long time. People worked hard to build mountain homes and roads. Today, many people enjoy mountain vacations. Mountains are great places to hike, hunt, swim, and ski!

GLOSSARY

chain: a group of related islands

cliff: a tall, steep area of land

elevation: the height something is above sea level

erosion: the wearing away of the earth by wind or water

lava: melted rock that comes to Earth's surface

peak: the highest point of a mountain

surface: the outer layer of something

FOR MORE INFORMATION

BOOKS

Amstutz, Lisa J. *Mountains.* North Mankato, MN: Capstone, 2020.

Tanumihardja, Pat. *Amazing Mountains Around the World.* North Mankato, MN: Capstone, 2019.

WEBSITES

Climbing Mount Everest

www.nationalgeographic.com/adventure/everest/reference/climbing-mount-everest/

Mount Everest is the tallest mountain in the world! This website from National Geographic has fascinating details about climbing Mount Everest.

Mountains Information and Facts

www.nationalgeographic.com/science/earth/surface-of-the-earth/mountains/

You can learn much more about mountains and how they form from this National Geographic website.

INDEX